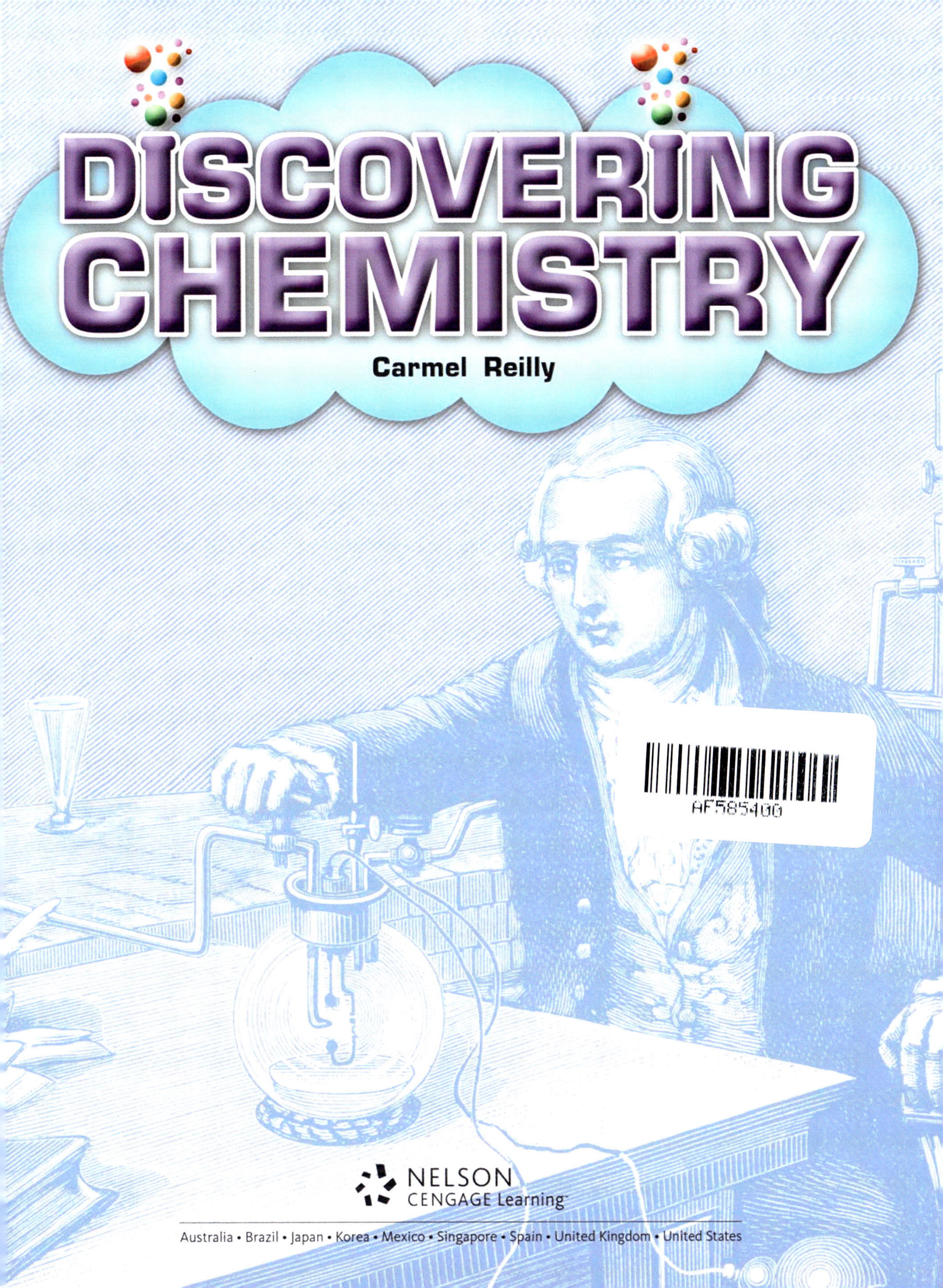

# DISCOVERING CHEMISTRY

Carmel Reilly



NELSON
CENGAGE Learning

Australia • Brazil • Japan • Korea • Mexico • Singapore • Spain • United Kingdom • United States

**Discovering Chemistry**

**Fast Forward**
**Silver Level 24**

Text: Carmel Reilly
Editor: Johanna Rohan
Design: Ami Sharpe
Series design: James Lowe
Production controller: Seona Galbally
Photo research: Fiona Smith
Audio recordings: Juliet Hill, Picture Start
Spoken by: Matthew King and Abbe Holmes
Reprint: Siew Han Ong

**Acknowledgements**
The author and publisher would like to acknowledge permission to reproduce material from the following sources: Photographs by Alamy/Brand X Pictures, p 14b/ Iconsight, p 8 inset/sciencephotos, p 13c/ Thinkstock, p 13b; Fotolia, back cover, pp 3, 5, 7b; iStockphoto, pp 4a-d, 12a-b, 23a-c; Mary Evans Picture Library, p 7a; Newspix/Gregg Porteous, p 22; Photolibrary/Carol and Mike Werner, p 18/ Lebrecht Music & Arts Photo Library, p 9/ Novosti Novosti, p 16/ Science Photo Library, pp 9 inset, 13a, 15a, 15c, 17/ Sheila Terry/SPL, cover, pp 1, 6, 14a/ STE, p 11/ The Bridgeman Art Library, p 8/ Science Photo Library, pp 10, 12c-d, 21b/ Andrew Lambert Photography, pp 15b, 15d/ E.R. Degginger, p 21a/ John Greim, p 5 inset.



ISBN 978 0 17 012710 3
ISBN 978 0 17 012705 9 (set)

**Cengage Learning Australia**
Level 7, 80 Dorcas Street
South Melbourne, Victoria Australia 3205
Phone: 1300 790 853

**Cengage Learning New Zealand**
Unit 4B Rosedale Office Park
331 Rosedale Road, Albany, North Shore NZ 0632
Phone: 0800 449 725

For learning solutions, visit **cengage.com.au**

Printed in Australia by Ligare Pty Ltd
6 7 8 9 10 11 20 19 18 17 16

Evaluated in independent research by staff from the Department of Language, Literacy and Arts Education at the University of Melbourne.

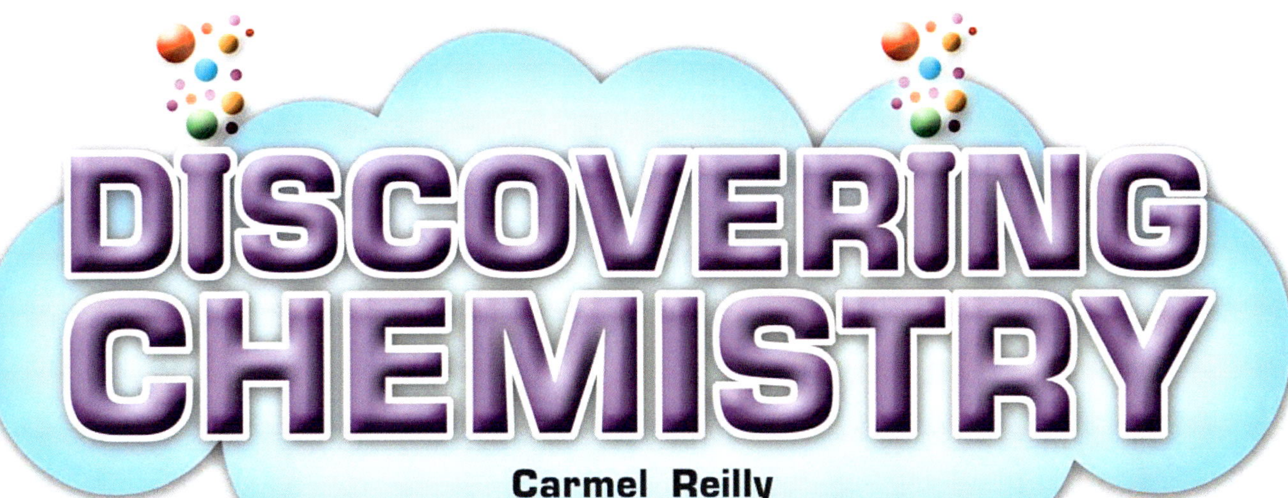

# DISCOVERING CHEMISTRY

Carmel Reilly

## Contents

Chapter 1

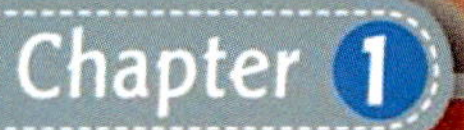

# INTRODUCING CHEMISTRY

Everyday things – all kinds of materials, as well as plant, animal and human life – are made up of a number of substances. These substances are called elements. Chemistry is the branch of science that studies the elements – what things are made of and how they can change.

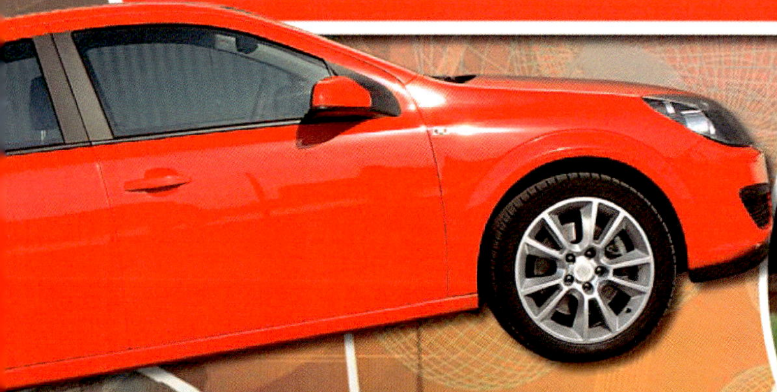

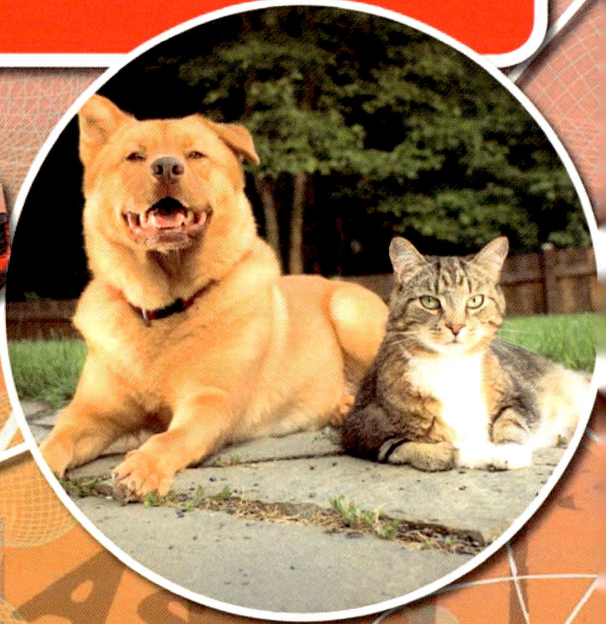

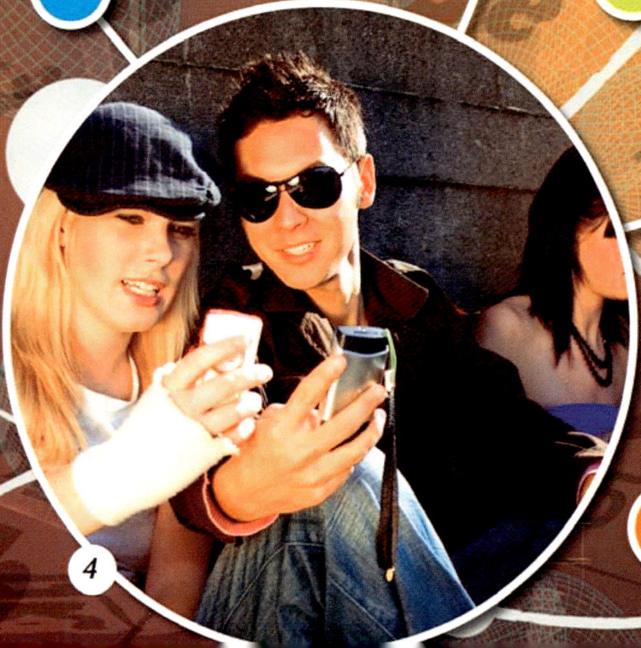

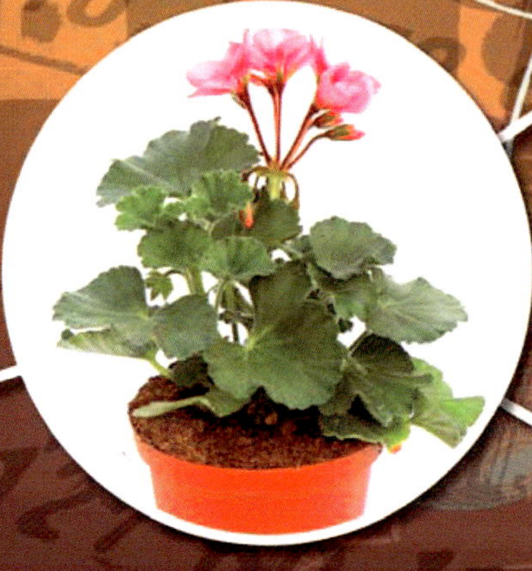

Scientists who study chemistry are called chemists. They **investigate** what makes up substances and how they react under different conditions. Chemists are also interested in finding ways to make new substances and materials.

# Chapter 2

# BEGINNINGS OF CHEMISTRY

For thousands of years, people have wondered what things are made from. People in ancient India and ancient Greece believed that everything was made from only four elements:

- air
- earth
- fire
- water.

men in ancient Greece

People in ancient China believed there were actually five elements:

- air
- earth
- fire
- water
- metal.

people in ancient China

People thought these things were elements because they seemed to be what all other materials and substances were made from. For example, pottery was made when clay from the earth and water were mixed and heated by fire.

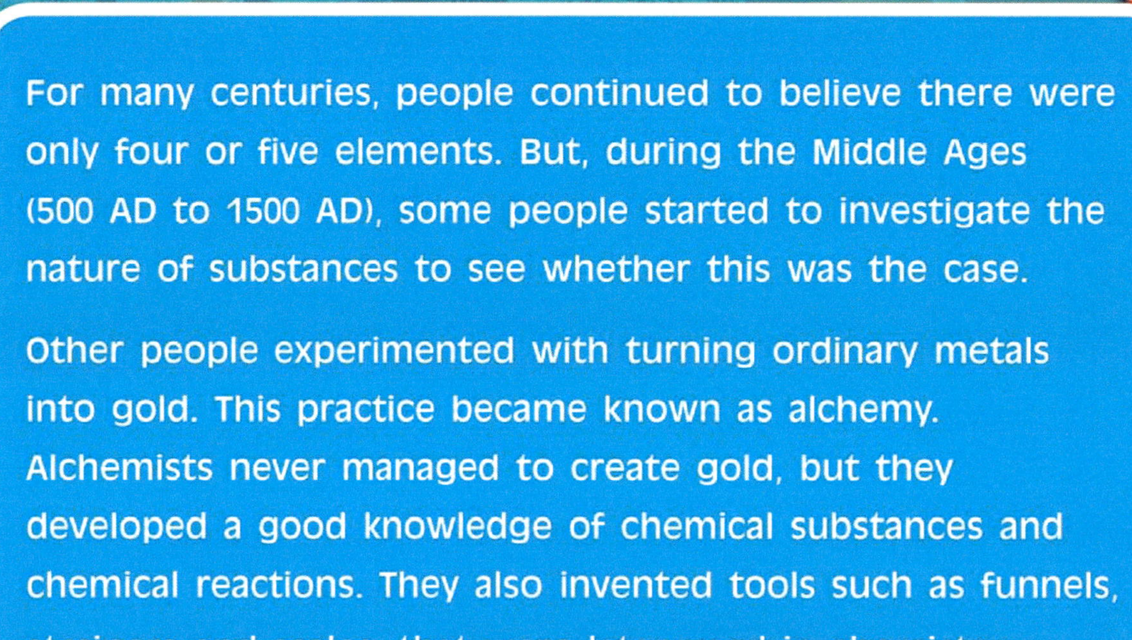

For many centuries, people continued to believe there were only four or five elements. But, during the Middle Ages (500 AD to 1500 AD), some people started to investigate the nature of substances to see whether this was the case.

Other people experimented with turning ordinary metals into gold. This practice became known as alchemy. Alchemists never managed to create gold, but they developed a good knowledge of chemical substances and chemical reactions. They also invented tools such as funnels, strainers and scales, that were later used in chemistry.

*scales*

*an alchemist*



Running Words 254

In the last 200 years, chemists have performed experiments that revealed that not everything is made only from the elements of air, earth, fire, water and metal.

Chapter 3

# THE BEGINNING OF MODERN CHEMISTRY

During the 1600s, scientists began to understand more about the make-up of materials and substances through experimenting. Scientists began to see that air, earth, fire and water were not elements themselves, but were made up of other elements.

Jean Baptiste van Helmont (1579–1644), a Belgian doctor and chemist

Robert Boyle

One of the first scientists to realise that there were many elements was Robert Boyle (1627–1691). Boyle is known today as the "father of modern chemistry". He believed that experiments and observations were the only way to find out about the world.

Boyle did many experiments with air. These experiments showed that air was not an element itself, but was made up of other smaller elements.

# Elements and Atoms

Everything is made from tiny particles called atoms. An element is a substance that is made of only one type of atom – a substance in its purest form. We now know that there are more than 100 elements, and about 80 per cent of them are metals, like gold, copper and iron.

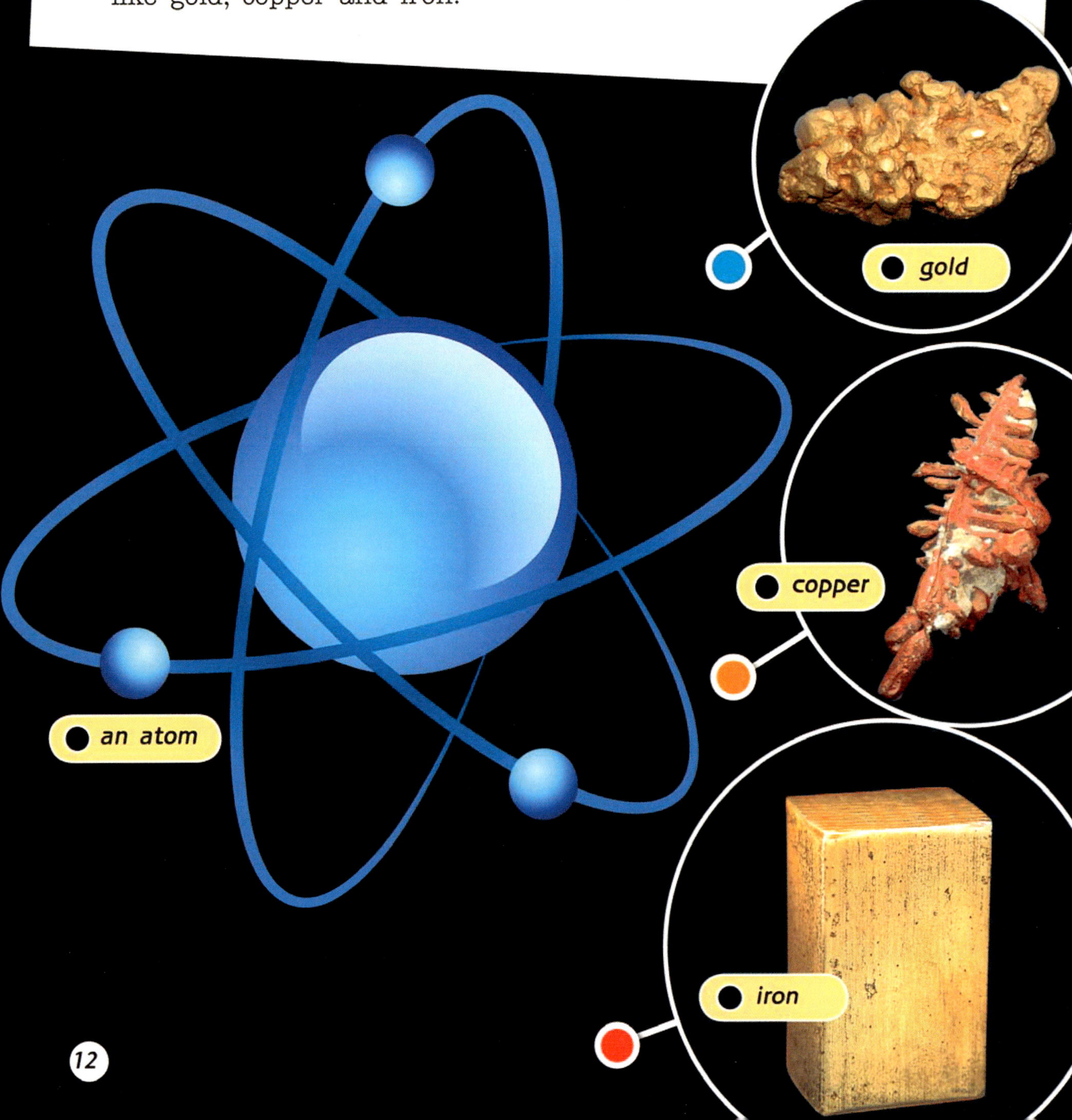

Henry Cavendish

Throughout the 1700s and 1800s, scientists made more advances in chemistry. Henry Cavendish (1731–1810) was able to make water when doing an experiment with air. From this he saw that water had to be made up of more than one element, and that water and air had elements in common. This experiment also led Cavendish to discover the element hydrogen. Around the same time, other chemists discovered oxygen and nitrogen.

A little later, the French chemist Antoine Lavoisier (1743–1794) was the first to show that air was a mixture of elements. He also proved that water is a compound of oxygen and hydrogen. Lavoisier wrote the first modern chemistry book and started the system of naming elements.

Humphry Davy (1778–1829) found a way to split compounds using an **electric current**. He also discovered a number of elements, including sodium and calcium. Around the same time, John Dalton (1766–1844) put forward the idea that everything is made of atoms, and that each element contains only one kind of atom.

## Compounds and Mixtures

A compound is a substance that is made up of more than one element joined together chemically. An example of a compound is pure water. It is made up of the elements hydrogen and oxygen.

A mixture is made up of more than one substance. An example of a mixture is air. It is made up of elements such as oxygen and compounds such as carbon dioxide.

# Chapter 4

# THE PERIODIC TABLE

By the 1860s, scientists had discovered 63 elements, each of which had been given an **atomic number**. A Russian scientist, Dmitry Mendeleyev (1843–1907), used these numbers to help design a table for the elements. He called this the periodic table. Making this table helped him and other scientists to understand the connections between each element, and to see how the elements could combine to make up different substances.

Dmitry Mendeleyev

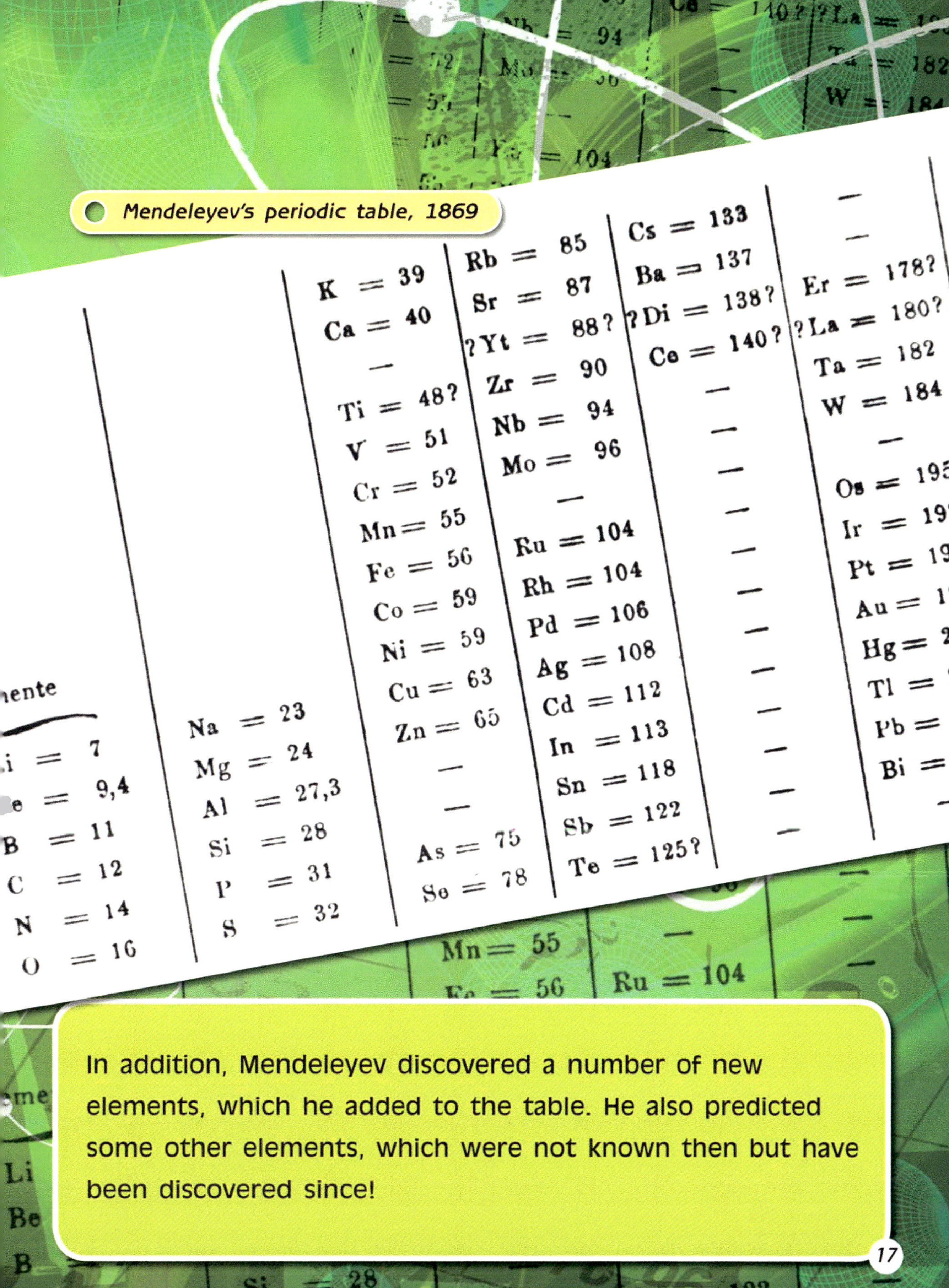

| ente | | | | | |
|---|---|---|---|---|---|
| | | K = 39 | Rb = 85 | Cs = 133 | — |
| | | Ca = 40 | Sr = 87 | Ba = 137 | — |
| | | — | ?Yt = 88? | ?Di = 138? | Er = 178? |
| | | Ti = 48? | Zr = 90 | Ce = 140? | ?La = 180? |
| | | V = 51 | Nb = 94 | — | Ta = 182 |
| | | Cr = 52 | Mo = 96 | — | W = 184 |
| | | Mn = 55 | — | — | — |
| | | Fe = 56 | Ru = 104 | — | Os = 195 |
| | | Co = 59 | Rh = 104 | — | Ir = 197 |
| | | Ni = 59 | Pd = 106 | — | Pt = 19 |
| i = 7 | Na = 23 | Cu = 63 | Ag = 108 | — | Au = 19 |
| e = 9,4 | Mg = 24 | Zn = 65 | Cd = 112 | — | Hg = 2 |
| B = 11 | Al = 27,3 | — | In = 113 | — | Tl = 2 |
| C = 12 | Si = 28 | — | Sn = 118 | — | Pb = |
| N = 14 | P = 31 | As = 75 | Sb = 122 | — | Bi = |
| O = 16 | S = 32 | Se = 78 | Te = 125? | — | — |

Mendeleyev's periodic table, 1869

In addition, Mendeleyev discovered a number of new elements, which he added to the table. He also predicted some other elements, which were not known then but have been discovered since!

The periodic table lists all the known elements. Some of these were discovered centuries ago, while others have only been discovered in the last few years.

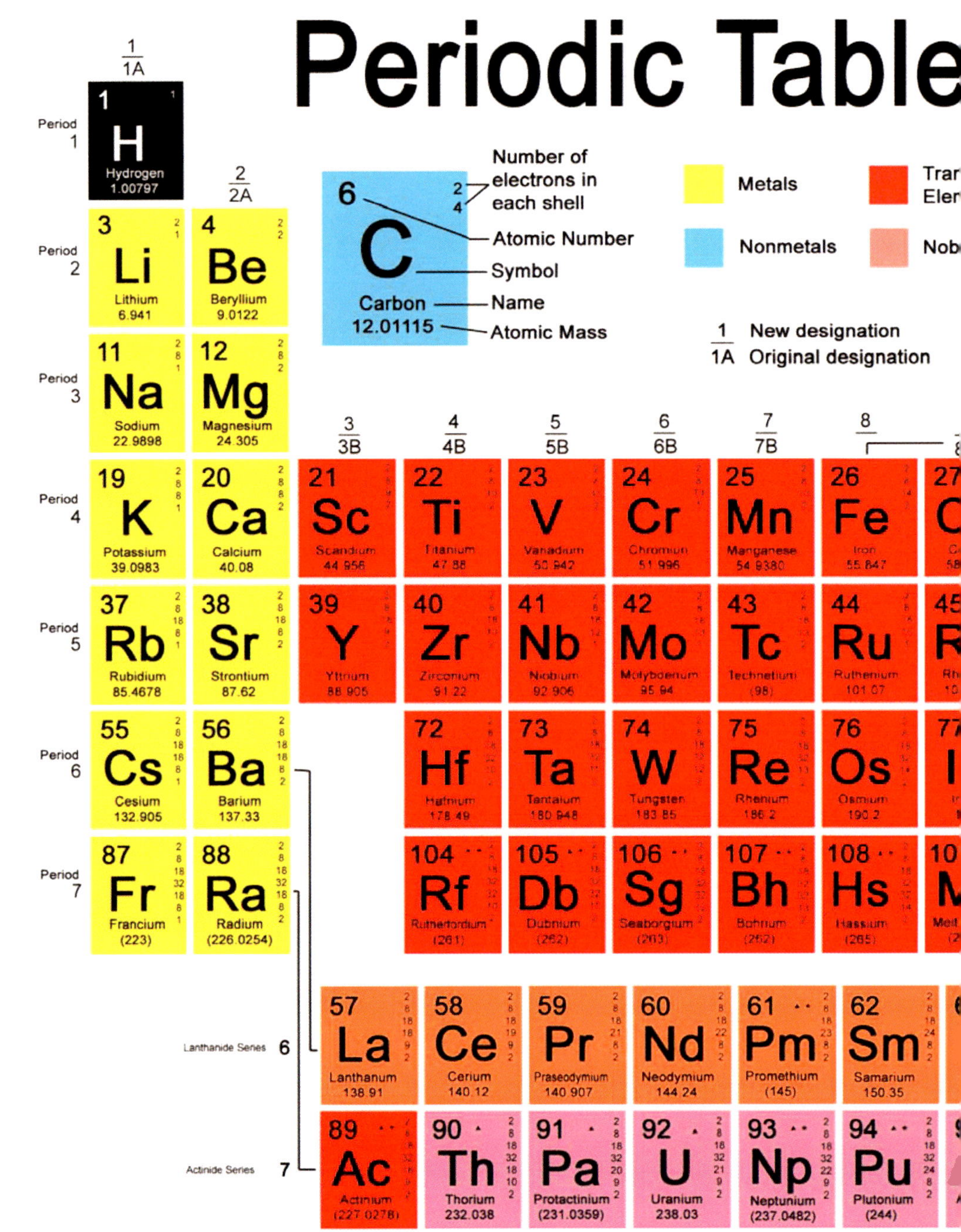

# of the Elements

Inner Transition Elements

* Synthetic

▲ Radioactive

( ) Atomic weight of most stable isotope

| 10 | 11 1B | 12 2B | 13 3A | 14 4A | 15 5A | 16 6A | 17 7A | 18 8A |
|---|---|---|---|---|---|---|---|---|
| | | | | | | | | 2 He Helium 4.0026 (2) |
| | | | 5 B Boron 10.811 (2, 3) | 6 C Carbon 12.01115 (2, 4) | 7 N Nitrogen 14.0067 (2, 5) | 8 O Oxygen 15.9994 (2, 6) | 9 F Fluorine 8.9984 (2, 7) | 10 Ne Neon 20.179 (2, 8) |
| | | | 13 Al Aluminum 26.9815 (2, 8, 3) | 14 Si Silicon 28.086 (2, 8, 4) | 15 P Phosphorus 30.9738 (2, 8, 5) | 16 S Sulfur 32.064 (2, 8, 6) | 17 Cl Chlorine 35.453 (2, 8, 7) | 18 Ar Argon 39.948 (2, 8, 8) |
| 28 Ni Nickel 58.69 | 29 Cu Copper 63.54 | 30 Zn Zinc 65.37 (2, 8, 18, 2) | 31 Ga Gallium 69.72 (2, 8, 18, 3) | 32 Ge Germanium 72.59 (2, 8, 18, 4) | 33 As Arsenic 74.9216 (2, 8, 18, 5) | 34 Se Selenium 78.96 (2, 8, 18, 6) | 35 Br Bromine 79.904 (2, 8, 18, 7) | 36 Kr Krypton 83.80 (2, 8, 18, 8) |
| 46 Pd Palladium 106.4 | 47 Ag Silver 107.868 | 48 Cd Cadmium 112.40 (2, 8, 18, 18, 2) | 49 In Indium 114.82 (2, 8, 18, 18, 3) | 50 Sn Tin 118.69 (2, 8, 18, 18, 4) | 51 Sb Antimony 121.75 (2, 8, 18, 18, 5) | 52 Te Tellurium 127.60 (2, 8, 18, 18, 6) | 53 I Iodine 126.9044 (2, 8, 18, 18, 7) | 54 Xe Xenon 131.29 (2, 8, 18, 18, 8) |
| 78 Pt Platinum 195.09 | 79 Au Gold 196.967 | 80 Hg Mercury 200.59 (2, 8, 18, 32, 18, 2) | 81 Tl Thallium 204.383 (2, 8, 18, 32, 18, 3) | 82 Pb Lead 207.19 (2, 8, 18, 32, 18, 4) | 83 Bi Bismuth 208.980 (2, 8, 18, 32, 18, 5) | 84 Po Polonium (209) (2, 8, 18, 32, 18, 6) | 85 At Astatine (210) (2, 8, 18, 32, 18, 7) | 86 Rn Radon (222) (2, 8, 18, 32, 18, 8) |
| 110 ** Ds Darmstadtium (269) | 111 ** Rg Roentgenium (272) | 112 ** Uub Ununbium (277) | 113 Uut | 114 Uuq | 115 Uup | 116 Uuh | 117 Uus | 118 Uuo |

Unknown elements 113 - 118 are shown in their predicted positions.

| 64 Gd Gadolinium 157.25 (2, 8, 18, 25, 9, 2) | 65 Tb Terbium 158.9254 (2, 8, 18, 27, 8, 2) | 66 Dy Dysprosium 162.50 (2, 8, 18, 28, 8, 2) | 67 Ho Holmium 164.930 (2, 8, 18, 29, 8, 2) | 68 Er Erbium 167.26 (2, 8, 18, 30, 8, 2) | 69 Tm Thulium 168.934 (2, 8, 18, 31, 8, 2) | 70 Yb Ytterbium 173.04 (2, 8, 18, 32, 8, 2) | 71 Lu Lutetium 174.97 (2, 8, 18, 32, 9, 2) |
|---|---|---|---|---|---|---|---|
| 96 ** Cm Curium (247) (2, 8, 18, 32, 25, 9, 2) | 97 ** Bk Berkelium (249) (2, 8, 18, 32, 26, 9, 2) | 98 ** Cf Californium (251) (2, 8, 18, 32, 28, 8, 2) | 99 ** Es Einsteinium (252) (2, 8, 18, 32, 29, 8, 2) | 100 ** Fm Fermium (257) (2, 8, 18, 32, 30, 8, 2) | 101 ** Md Mendelevium (258) (2, 8, 18, 32, 31, 8, 2) | 102 ** No Nobelium (259) (2, 8, 18, 32, 32, 8, 2) | 103 ** Lr Lawrencium (260) (2, 8, 18, 32, 32, 9, 2) |

# Chapter 5

# SOME OF THE ELEMENTS

| Iron | Platinum | Phosphorus |
|---|---|---|
| symbol: Fe | symbol: Pt | symbol: P |
| Discovered in ancient times | Discovered in the **pre-Columbian** era | Discovered in 1669 by Hennig Brand |
| Used for tools, ornaments and weapons. Now used in car manufacturing and structural parts for buildings. | Used in jewellery, laboratory equipment, and dentistry | Used in detergents and fertilisers |

Fe Pt P

| Hydrogen | Radon | Plutonium |
|---|---|---|
| symbol: H | symbol: Rn | symbol: Pu |
| Discovered in 1766 by Henry Cavendish | Discovered in 1898 by Friedrich Ernst Dorn | Discovered in 1940 by G.T. Seaborg |
| Used in balloons and metal refining | Used in the treatment of cancer | Used in bombs and nuclear reactors |

H Rn Pu

# Chapter 6

# CHEMISTRY TODAY

Without chemistry and all the chemists who have investigated and created new substances, the world today would be a very different place.

By understanding how elements combine and what substances are made of, chemists have been able to work out how to make thousands of new and useful products, such as synthetic fabrics, LCD and plasma computer and TV screens, paints, sun screen, cosmetics and medicines.

## Glossary

**atomic number** a number that identifies an element. It is the number of protons found in the nucleus of an atom.

**electric current** a flow of electricity

**investigate** to carry out research into something

**pre-Columbian** the era before and during Christopher Columbus' first landing in the Americas, in 1492

## Index